SPARK ISLAND

KS2 Skills Practice
SCIENCE
Learning Adventures

Simon Greaves

Introduction

About this book

Science affects our everyday lives: the way our body works, how plants grow, the materials we use to make things, how we use light and electricity and what causes day and night. Like the Elders of Sparkopolis, as they design and improve their amazing capital city, we need to understand and use scientific and technical skills. So why not join The Gang to find out more about Science and how exciting it can be!

How to use this book

Think about the best time for using this book. It might be easier at a weekend or early in the evening. Most of all, pick a quiet time when your child is eager to learn and not too tired. Find a suitable place where he or she can work comfortably without being disturbed, then make a start on one of the activities.

Your child may prefer to work through the book page by page, or alternatively you could suggest activities that you feel will be more useful. Whichever approach you adopt, try to make it an enjoyable and positive experience for your child. Discuss the activities together and give lots of praise and encouragement along the way. All the answers are at the back of the book for quick and easy checking!

How you can help

- Remember little and often is best! Children's brains can only take so much, it's best to stop before they get bored and grumpy.
- You can help your child by reading questions, testing knowledge of facts, marking answers and discussing topics that your child finds difficult.
- Use the *Spark Island KS2 National Tests Science* for some more exciting activities, or log on to the Spark Island website (www.sparkisland.com) for your child to take part in some interactive fun.
- Most of all encourage your child, praise his or her efforts and use the gold stickers to reward good work.

Contents

Introduction	2
Living things and classification	4
Healthy lifestyle	6
Teeth	10
The human body and life cycle	12
Green plants	14
Living things in their environments	18
Micro-organisms	22
Properties of materials	23
Reversible and non-reversible changes	26
Separating mixtures	30
The Earth and beyond	32
Forces	36
Electricity	40
Light	42
Sound	44
Answers	46

Living things and classification

There are many different life forms on Spark Island, each with their own characteristics: the short stature of the Spironauts, the extraordinary mental powers of the Phliplids and the eight legs of the Spydrax. But like all living things in our world, they have the same seven life processes.

1 The seven life processes

Fit the words into the grid so that they spell out the seventh life process in the shaded column.

FEED
MOVE
GROW
SEVEN
SENSITIVE
REPRODUCE
EXCRETE

Zeb's fact file

A blue whale's tongue is about the same size as a fully grown African elephant!

2 The five senses

(a) **Taste** is commonly believed to be one of the five senses. Can you find out the other four?

TONGUE TWISTER

How much wood could a woodchuck cut
If a woodchuck could cut wood?
He'd cut as much wood as a woodchuck could
If a woodchuck could cut wood.

(b) Different parts of your tongue sense different flavours.

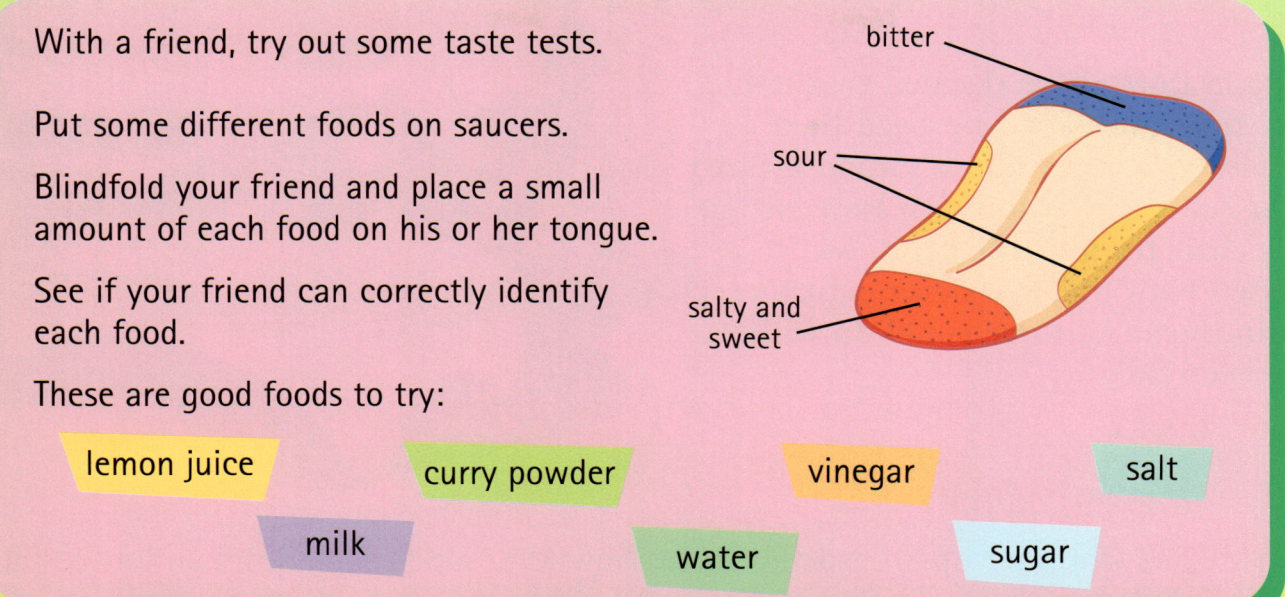

With a friend, try out some taste tests.

Put some different foods on saucers.

Blindfold your friend and place a small amount of each food on his or her tongue.

See if your friend can correctly identify each food.

These are good foods to try:

lemon juice   curry powder   vinegar   salt
milk   water   sugar

3 Classifying using a key

Many animals and plants belong to the same families and therefore have similar features, but how do you work out which one is which? The answer is to use a key.

Use the key below to work out what type of creature is shown in the picture.

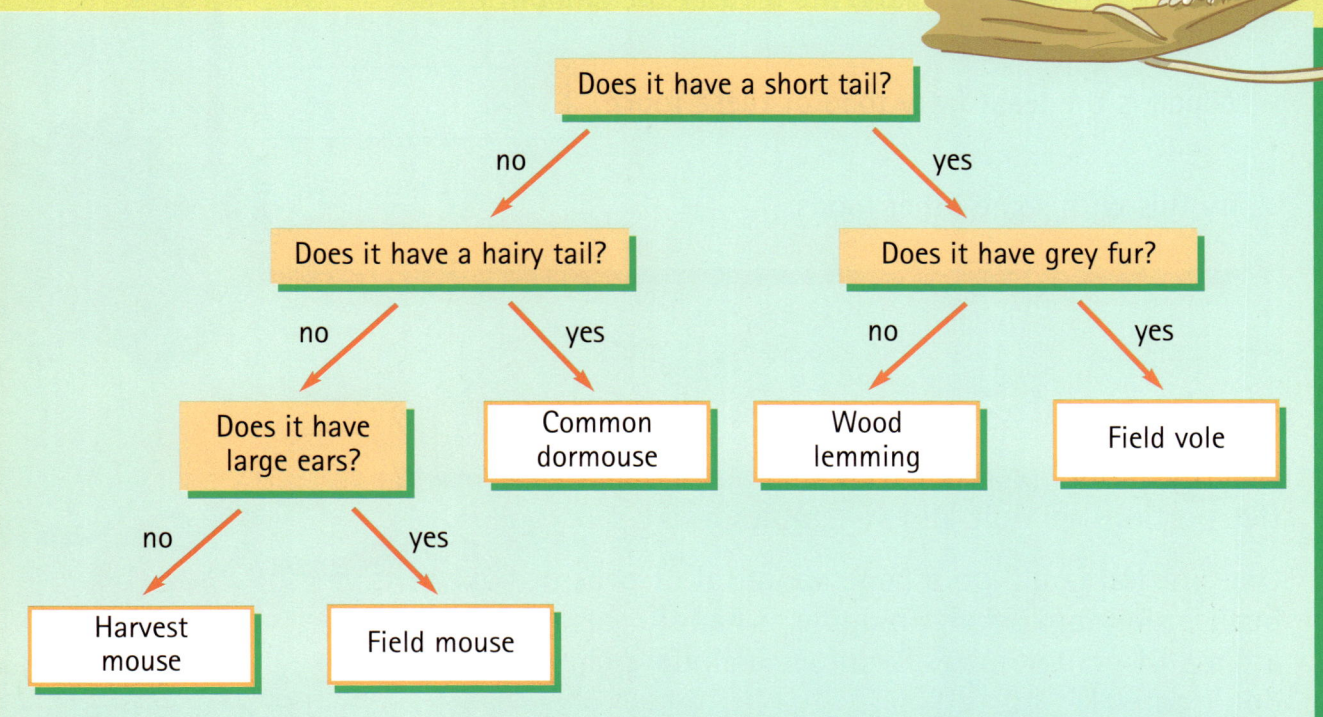

Healthy lifestyle

Nina and Dotty have decided that it's about time the Gang improved their lifestyle. Milo is unhealthily fond of fatty foods. Zeb spends too much time sitting in his chair reading and Strat, whilst obsessed with trainers, never actually runs in them. All three could do with some exercise.

1 Finding fatty foods

Nina and Dotty set up an experiment to show Milo which foods contain lots of fat. Try it for yourself.

You will need some pieces of filter paper or kitchen paper, a couple of crisps, and some small pieces of food: brown bread, butter and chocolate. Crush the crisps. Then cut a small piece of each food to about the size of a peanut. Place each piece on a separate piece of filter paper. Leave them for six hours.

The fat in each food is absorbed by the paper, leaving a greasy mark. The bigger the mark, the greater the amount of fat.

Which is the least fatty food of these four? _____

Try this with some other foods.

2 Five fruit and veg

Healthy eating experts tell us that we should eat at least five portions of fruit and vegetables a day. Do you?

One portion could be a large spoonful of peas or baked beans, a tomato salad, an apple, a handful of grapes or a glass of orange juice. Potatoes, including chips, don't count!

For one week record all the fruit and vegetables you eat each day.
Work out whether you have eaten at least five portions a day.

Monday	Tuesday	Wednesday	Thursday	Friday	Saturday	Sunday

3 A balanced meal

The picture shows a balanced meal. This means it includes foods from all the food groups and doesn't have too many sugary or fatty foods.

Complete the labels to show the main nutrients each item provides.

Zeb's food facts

A healthy diet includes foods of different types.

Carbohydrates provide energy: bread, pasta, potatoes, cereal and biscuits.

Proteins help growth and repair: meat, fish, eggs, beans and cheese.

Fats provide stored energy: cheese, butter and oil.

Fibre helps digestion: bran, wholemeal bread, fruit and vegetables.

Vitamins and **minerals** are needed for good general health: many foods especially cereal, dairy products, fruit and vegetables.

4 Fruit kebabs

Nina and Dotty have created a healthy snack for the Gang. Here is the recipe for you to try.

You will need some wooden skewers or cocktail sticks and a knife. Choose some colourful fresh fruits such as strawberries, apple, grapes, kiwi fruit, banana and melon, or tinned fruits such as peaches or pineapple. Carefully cut the fruit into chunks (about 2 cm cubes) and push a variety of fruit chunks on to each skewer to make colourful kebabs.

5 Keeping healthy

Exercise is extremely important for keeping our bodies working well. It makes muscles strong, keeps the lungs and heart working properly and helps control weight.

When the heart beats, it pumps blood around the body. This can be felt as a 'pulse'. You can feel your pulse at different places on your body, such as your wrist, neck or chest. Pulse is measured as a number of beats per minute.

DID YOU KNOW?
- In the UK about 15% of all school children are overweight.
- On average, people in the UK only eat three portions of fruit and vegetables a day. This is not enough!

How to measure your pulse
Place your first two fingers on your neck just under your jaw. Wait until you can feel your pulse.
Now count the number of beats for ten seconds. Multiply this number by six to give you the number of beats per minute. You will find it easier if someone times the ten seconds for you.

Exercise increases your pulse rate because your heart is having to work harder to keep up with the extra activity. Regular exercise helps to make your heart stronger. After all, it is a muscle.

Investigate how your pulse changes with activity. For each activity listed below, measure your pulse (as described above) immediately after.

Activity	Pulse (beats per minute)
Sitting still in a chair for 5 minutes	
Walking briskly for 5 minutes	
Jumping or skipping for 1 minute	
Running for 1 minute	

6 Drugs

Drugs are harmful if they are misused. Many have damaging effects on the body and mind.

Draw lines to match the four habits to the effects they have on the body.

alcohol abuse

smoking

sniffing solvents

taking drugs

damages nose

heart problems

slows reactions

brain damage

addictive

risk of overdose

lung cancer

liver damage

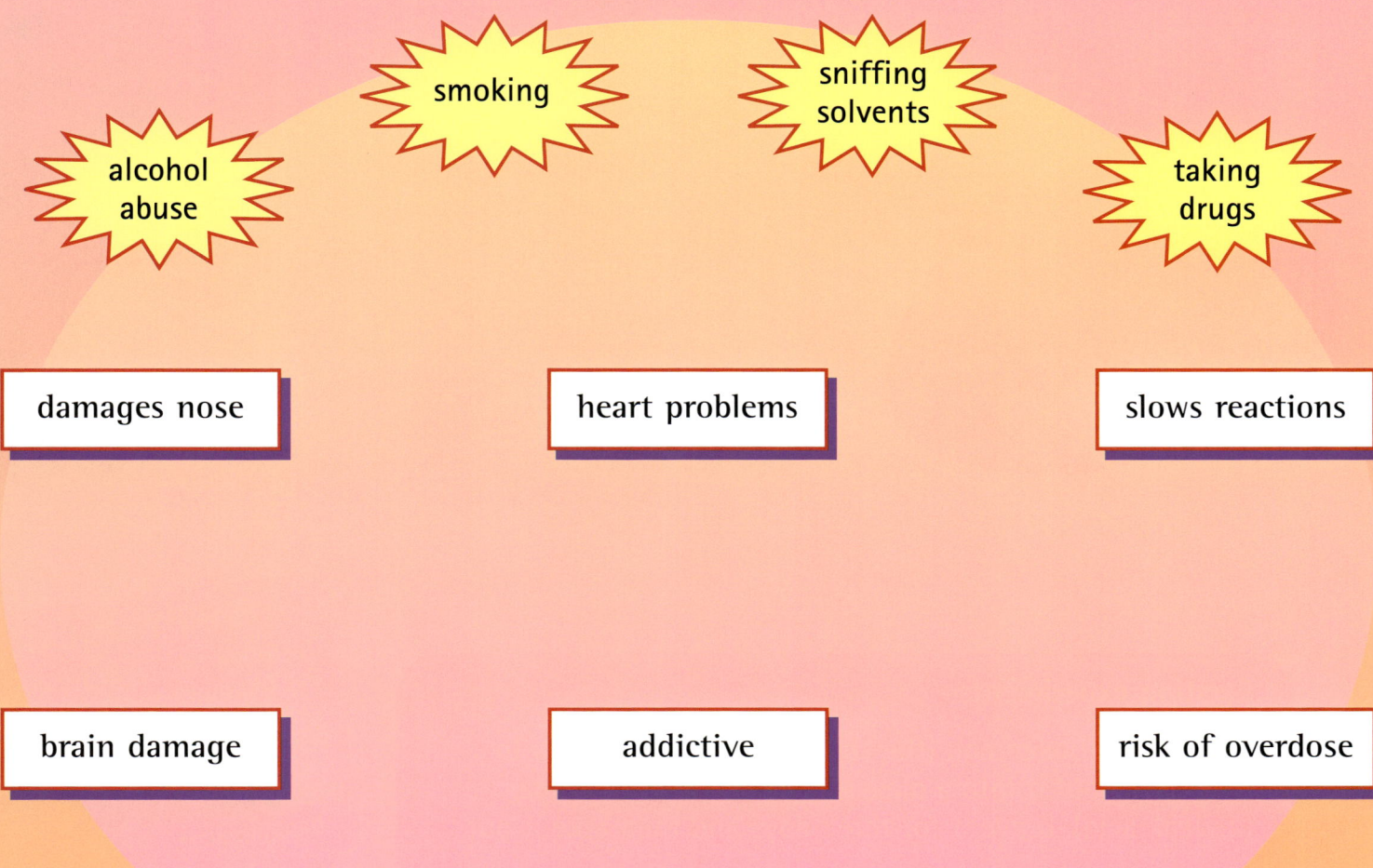

Misuse of drugs can destroy lives. However, some drugs are helpful. Doctors prescribe drugs for many good reasons such as fighting infections, reducing pain and preventing and treating fatal diseases.

IN THE PINK

A healthy lifestyle is the way
To keep a heart attack at bay.
Do not smoke and watch the drink
To keep your body in the pink.

Teeth

Milo is very worried as one of his milk teeth has fallen out. Zeb tries to comfort Milo by explaining that his milk tooth will be replaced by an adult tooth.

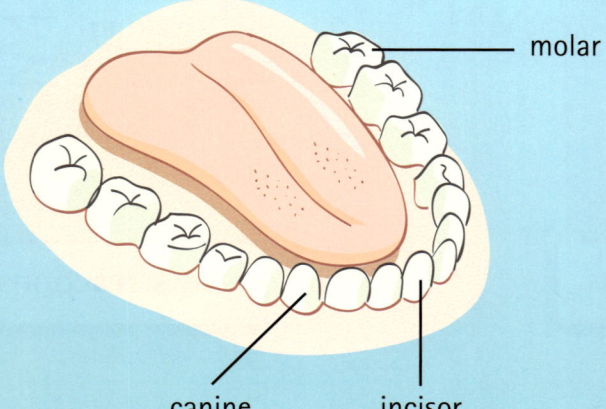

Zeb's teeth facts

An adult has 32 teeth.

There are three types of teeth: **incisor**, **canine** and **molar**.

An incisor cuts and snips food.

A canine pulls and rips food.

A molar chews and grinds food.

Plaque is a sticky coating which is formed on your teeth after eating.

Bacteria in the plaque produces an acid which attacks your teeth.

1 Exciting experiments

(a) Take a hard boiled egg (the shell is a similar material to our teeth), put it into a glass and cover it with vinegar (a weak acid). Leave it for three days and then take it out.

What has happened?

(b) Teeth can be stained by some foods and drinks so it's important to clean them regularly.

Put another hard boiled egg into a glass of undiluted blackcurrant squash. Leave it for a few days and take it out.

What has happened?

A HAPPY MOLAR

Teeth are fab, teeth are cool,
Eat too much sugar and you'll be a fool.
Brush them daily, drink less cola,
Follow these rules for a happy molar!

DID YOU KNOW?

- A snail's mouth is really small but contains over 25,600 teeth.
- Human teeth are almost as hard as rock.

2

Make a toothbrush holder

You will need either a 2 pint or 4 pint plastic milk container.

1 Cut the milk container as shown.

> **Top tips for healthy teeth**
> Avoid sugary foods.
> Brush your teeth twice a day.
> Visit your dentist regularly.
> Floss your teeth.

2 Copy out 'Top tips for healthy teeth' (see box above) on to pieces of coloured paper.

3 Cut out each Top Tip and stick them around the sides of the base of the container.

4 Invert the top and place it in the base.

5 Put your toothbrush into the small hole and your toothpaste and floss in the big one.

6 Use twice a day!

TWICE

Brush your teeth twice a day,
If you don't there'll be a price to pay.
Visit your dentist twice a year,
To make sure that no cavities appear.

TOOTHY TALES

Nina loves chocolate! But she should remember to brush her teeth regularly.

Xybok are herbivores, like cows and rabbits. They have long incisors to snip and cut plants and very flat molars to crush and grind their food, especially raspberries and gooseberries! Carnivores, on the other hand, have long canines to kill and rip their prey.

The human body and life cycle

Sam Spironaut has been hurt during a Xybok race at the stadium. He is worried that he has broken one of his four arms. Another Spironaut takes him to the Sparkopolis hospital to have it X-rayed. The X-ray reveals a small fracture below the elbow, so Sam has to wear his arm in a sling.

1 The skeleton

Across
2 Another name for the backbone
4 A mineral needed for healthy bones
5 A broken bone
7 One of the jobs of the skeleton
8 One of the organs protected by the rib cage

Down
1 A type of joint
2 A part of the skeleton that protects the brain
3 Attached to the bones to help movement
5 The longest bone in the body

2 The ball and socket joint

Bones in the skeleton are jointed in different ways. One type of joint is the ball and socket. Your shoulders and hips have this type of joint.

Make this model to study how joints like these work.

You will need an egg cup, a table tennis ball, some modelling clay and a pencil. Rub the inside of the egg cup with some oil to act as a lubricant.

Attach the pencil to the ball with some modelling clay.

Place the ball in the egg cup and move the pencil. Study the different directions in which the ball can move.

3 Lung capacity

Zeb has designed an experiment to test how much air each member of the Gang can hold in his or her lungs.

Try this experiment at home.

You will need an empty 2 litre water bottle, a long bendy straw or a piece of plastic tubing, a washing up bowl and someone to help you!

Half-fill the washing up bowl with water. Fill the plastic bottle with water, cover the end with your hand and turn it upside-down with the neck under the water in the washing up bowl. Remove your hand. The water in the bottle will not flow out!

Push the straw or tube into the neck of the bottle under the water. You will need someone to hold the bottle for you.

Now take a deep breath and blow through the straw.

The air from your lungs will push the water out of the bottle and into the bowl.

Mark the water level on the bottle.

Try this experiment with some of your friends and see who has the biggest lungs.

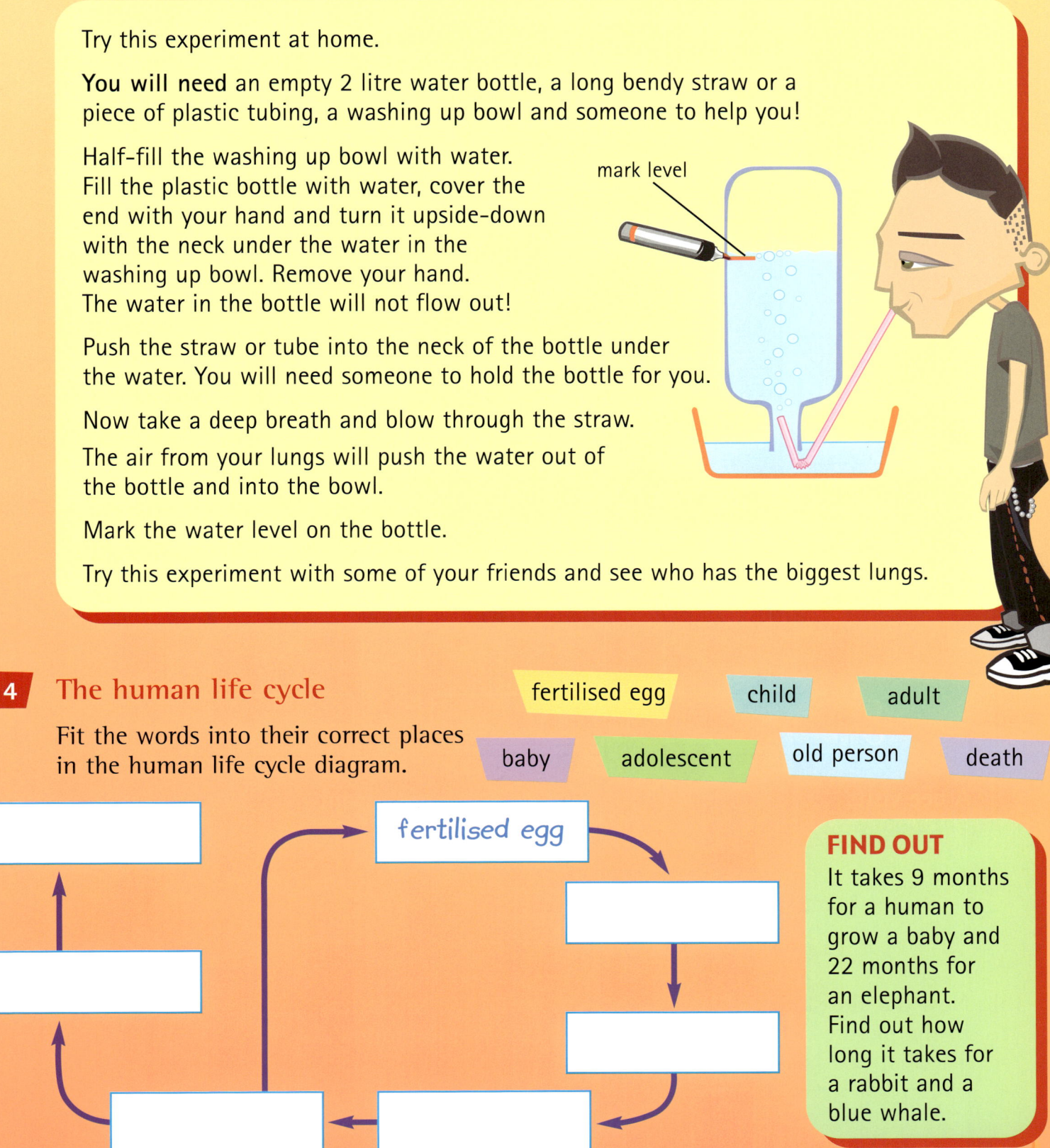

mark level

4 The human life cycle

Fit the words into their correct places in the human life cycle diagram.

fertilised egg   child   adult   baby   adolescent   old person   death

fertilised egg

FIND OUT

It takes 9 months for a human to grow a baby and 22 months for an elephant. Find out how long it takes for a rabbit and a blue whale.

Green plants

The Gang have been on a visit to the countryside outside Sparkopolis. They were fascinated by the many different plants and flowers they saw. Milo was so interested he wants to learn more about plants. Zeb has come up with some activities for him to do.

1 The life cycle of a poppy

(a) Each picture shows a stage in the life cycle of a poppy. Match the words to the pictures. Label each picture.

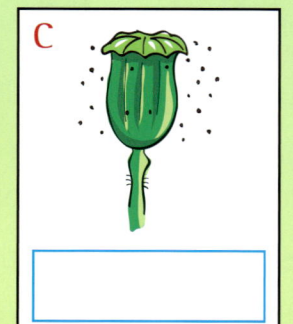

A bee visits the poppy.   The seed falls to the ground.   A small plant grows.   The wind scatters the seeds.

The plant flowers.   The seed germinates.   The flower dies leaving a seed head.

(b) Now put the stages of the life cycle in order. Use the letters in the pictures, starting with 'The seed falls to the ground'. The first one has been done for you.

A

2 What plants need

Plants only grow well if they have the right conditions. Two factors which affect plant growth are the amount of light and water the plant receives.

Try this experiment using cress seeds to study these factors.

You will need a packet of cress seeds, some cotton wool and 3 saucers. Lay the cotton wool in each saucer and sprinkle some seeds on to the surface. Set up the conditions for each saucer as shown in the table and record your observations over two weeks. Predict which seeds will grow best.

Saucer	Conditions	Observations	
		end of week 1	end of week 2
1	Dry cotton wool on window-sill		
2	Wet cotton wool on window-sill		
3	Wet cotton wool in dark cupboard		

3 The parts of a flower

Strat bought some flowers for Nina. Unfortunately, before he could give them to her, Zeb found they were perfect for showing Milo the parts of a flower - and dissected them!

Label the **stamen, stigma** and **ovary** on the flowers below.

4 Plant stems

In a green plant the stem transports water to other parts of the plant.

A good way to see this is to stand a tulip in a jar of coloured water or ink for a few hours. Then make a vertical cut in the tulip stem to check how far the colouring has reached. This shows how far water would be carried through a plant.

5 Plants as food

We eat a variety of plants but, depending on the plant, we eat different parts of it.

Which part do we eat?
Choose one of these words to fill the spaces.

fruit   flower   leaves   stem   root

(a) broccoli _____

(b) tomato   _____

(c) celery   _____

(d) carrot   _____

(e) cabbage   _____

(f) parsnip   _____

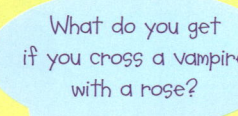

What do you get if you cross a vampire with a rose?

A flower that goes for your throat when you sniff it!

Zeb's fact file

One of the biggest sunflowers ever grown was over 8 metres tall.

One of the heaviest carrots ever grown weighed over 10 kg.

6 Seed dispersal

Zeb shows Milo how to blow a dandelion clock and explains that one way seeds can be dispersed is by the wind. He explains that seeds are scattered in different ways.

Complete the sentences about each type of seed using the words and phrases given below.

animals   hooks   pods   eaten by an animal   explode   the wind

A The seed has wings which help it to be dispersed by _____.

B The seed is _____ and then excreted as droppings.

C The seeds are stored in _____ which _____ to scatter the seeds.

D The seed has burrs which act like _____ and attaches itself to _____.

Living things in their environments

All animals and plants prefer to live in suitable environments to which they are specially adapted. Xybok are docile grazing animals which live on the quiet plains outside Sparkopolis and feed on fresh gooseberries and raspberries.

1 Woodlice habitat

(a) Try this experiment.

You will need a shallow cardboard box divided into four sections. Set up each section as shown in the diagram.

Spray these 2 sections with water until the cardboard is damp.

Cover these 2 sections with black card or newspaper.

Leave these 2 sections open to daylight.

dark damp	dark dry
light damp	light dry

You will need to collect 10 woodlice to use in the experiment. Woodlice live in gardens and woodland and are often to be found under stones. Collect them carefully in a jar. Use a piece of card to ease them from the ground and into the jar. Carefully place the woodlice in the light, dry section in the box. Wait a while and watch what happens.

(b) Which two variables are being investigated in this experiment?

(c) Write down what you see.

(d) What do you think this experiment tells you about the habitat preferred by woodlice?

Zeb's fact file

Producer: a plant that makes its own food.

Consumer: an animal that eats food.

Prey: an animal that is hunted.

Predator: an animal that hunts other animals for food.

REMEMBER
When you have finished the experiment, return the woodlice to the place you found them.

2 Where do I live?

Look at the pictures of the three different environments. Match these plants and animals to their correct environment.

fern, squirrel, frog, kelp, reeds, crab, puffin, heron, badger

Forest

1 _____

2 _____

3 _____

Pond

1 _____

2 _____

3 _____

Seashore

1 _____

2 _____

3 _____

3 Worms

The worms of Sparkopolis have a habit of popping up all over the place, sometimes when least expected!

We hardly ever see our own earthworms unless it is raining or the soil is being disturbed. They are obviously best suited to life underground.

Try these earthworm investigations.

(a) **Do worms make a noise?**
Carefully collect an earthworm from the garden and put it on a white sheet of paper. Put your ear close to the paper and listen. You should hear a scratching noise. This is because a worm has tiny bristles on its body which help to anchor it in the soil and move across the surface.

(b) **What conditions do worms prefer?**
Now place a damp piece of paper next to the dry sheet. Leave a 5 cm gap between the two sheets. Watch how the worm moves from the dry to the damp piece. Worms prefer damp conditions.

How can you tell which end of a worm is which?

Tickle the middle and see which end laughs!

(c) Make a wormery

Don't forget to put your worms back in the soil.

You will need a large, clean glass or plastic jar, leaves and grass, soil, sand, coloured aquarium gravel and 3 earthworms carefully collected from outside.

Fill the jar with layers of leaves and grass, soil, sand and gravel, as shown and pour some water in to moisten it. Don't make it too soggy. Place the worms on the surface and watch them burrowing down through the layers, pulling the leaves and grass with them.

Watch your worms' activities for about a week, making sure you keep the wormery moist.

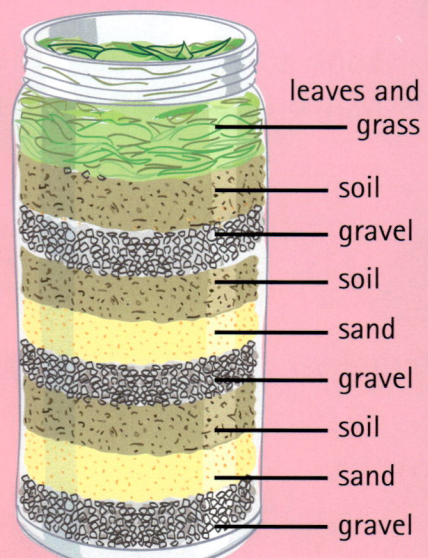

leaves and grass
soil
gravel
soil
sand
gravel
soil
sand
gravel

4 Food chains

Most plants and animals feature in different food chains. Most food chains start with a producer, usually a green plant.

Write these words into the boxes to make three food chains.

sparrow, caterpillar, heron, grass, kestrel, ladybird, cow, frog, human, greenfly, rose, lettuce

5 A wildlife garden game (for 2 players)

You will need 2 counters and 1 dice.

Take turns to roll the dice, moving the number of squares given on the dice. If you land on a green square you can move up the arrow because you have done something good for the environment. If you land on a red square you must move down the arrow because you have done something damaging to the environment. You must finish exactly on the last square.

WIN 64	You've destroyed a spider's web. 63	62	61	60	59	58	57
49	You've forgotten to put food on the bird table. 50	51	52	53	54	You've caught a butterfly and broken its wing. 55	56
Birds have built a nest in the nesting box. 48	47	46	45	Your cat has worked out how to climb on to the bird table. 44	43	42	41
33	34	You've set up a bird table with seeds. 35	36	37	38	You've planted a butterfly bush. 39	40
32	You've put up a nesting box. 31	30	29	28	You've let your pond freeze over. 27	26	25
You've created a mini meadow. 17	18	Frogs have laid spawn in your pond. 19	20	21	22	You've forgotten to put plants in the pond. 23	24
16	You've disturbed a hibernating hedgehog. 15	14	13	12	11	10	9
START 1	2	3	4	You've built a pond. 5	6	7	8

Micro-organisms

The Elders enjoy an occasional drink of raspberry wine and are aware of the helpful nature of the micro-organisms in yeast. Without yeast the wine would not ferment.

1 Helpful and harmful micro-organisms

Carefully examine the picture below. Put a circle around the **three** helpful micro-organisms at work. Put a cross over the **three** examples of harmful micro-organisms.

HYGIENE HINT
Always wash your hands before touching food.

2 Decaying materials

Compost is made from decaying rubbish. Micro-organisms feed on the rubbish and cause the materials in it to decay.

Circle the materials that can be broken down by micro-organisms.

| glass bottle | newspaper | wool socks | brick |
| CD | apple | metal spoon | leaves |

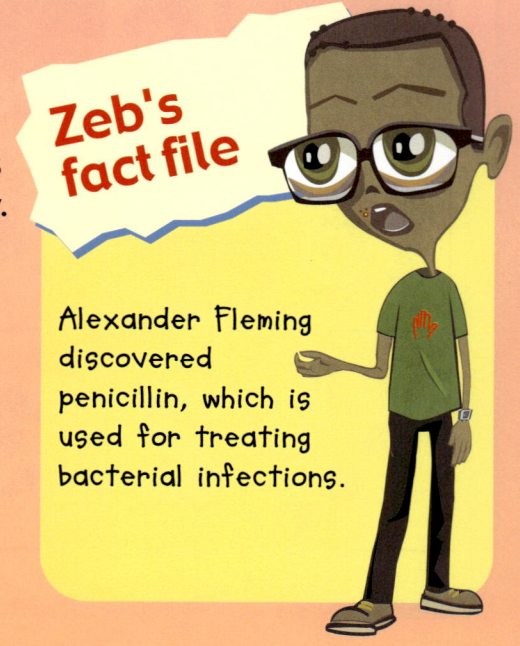

Zeb's fact file

Alexander Fleming discovered penicillin, which is used for treating bacterial infections.

Properties of materials

Fashion-conscious Dotty has been designing some clothes for the Gang to suit different situations. She has chosen certain fabrics because of their special properties.

1 Material properties

Different materials have different properties. Look at the table below. Put a tick in each column to show the properties of each of the four materials.

	magnetic	non-magnetic	man-made	natural	insulator	conductor
wood						
iron						
aluminium						
plastic						

2 Thermal insulators and conductors

Strat is washing the dishes after the Gang's dinner. He leaves three spoons to soak in the hot water. One spoon is made of steel, one is made of plastic and the other is made of wood. When he takes out the spoons he notices that one of them is very hot and the other two are just warm.

(a) Which spoon is hot?

(b) Explain your answer.

3 Making plastic

(a) Plastic is a material which has many uses. Take a look around your kitchen and make a list of all the items that are made from plastic.

Handle borax carefully. Wash your hands after using it.

(b) Try making your own plastic using this method.

> Stir 250 ml of warm water into a bowl containing two tablespoons of borax (this is a cleaning agent which can be bought from a chemist). When it has completely dissolved, leave it to cool thoroughly. Put about four tablespoons of PVA glue (you can buy this from a stationery shop) into a bowl and mix in 125 ml of warm water. Then gradually add the borax solution.
>
> You will notice that the mixture of the two liquids has turned into a rubbery plastic. Try stretching it and rolling into a ball. It should bounce!

Zeb's fact file

A bottle of lemonade is a good example of something that includes all three material states. The plastic bottle and cap are both **solids**. The lemonade is **liquid**. The bubbles are **gas** (carbon dioxide).

4 Solid, liquid and gas

In the previous experiment two liquids were mixed to form a solid. This experiment involves materials in solid, liquid and gas form.

Put a tablespoon of bicarbonate of soda into a cup and pour a teaspoon of vinegar on to it. Then complete the following using the words **solid**, **liquid** and **gas**.

	+		→		+	

5 Are bricks waterproof?

Find a brick. Weigh it on some kitchen scales and write down its weight. Put the brick in a bucket of water for a few hours. Now weigh the brick again and write down its weight.

You should find that the brick weighs more when wet than when dry. This is because the brick is made from a **permeable** material: it allows water into it. Some of the water in the bucket has been soaked up by the brick.

DID YOU KNOW?
Diamond is the hardest known substance. It is made of carbon: the same element as coal!

Alum Bay on the Isle of Wight is famous for its sand, which is found in many different colours.

6 Soils

Nina is growing some plants in small pots. She is not sure which type of soil she should grow them in so she tries out three different mixtures. After a couple of weeks of regular watering she notices that one plant is thriving, one is soggy and rotting and the other is dry and shrivelled up.

Match each type of soil to the correct picture of the plant.

very sandy soil

heavy and sticky clay soil

soil with some gravel and sand in it

A

B

C

7 A structure challenge

Some materials can be made stronger by changing their shape, for example paper. A sheet of paper is easily torn and is not very strong but it can be folded into different shapes which increase its strength. For example, folding the paper into a cylinder makes it a lot stronger.

Try this!
Make a structure which can cross a gap of 50 cm and support one heavy book (like a dictionary or phone book). You can only use 1 A4 sized sheet of thin card, 2 A4 sheets of paper, 2 sheets of newspaper, 10 paperclips, 15 small pieces of sticky tape and a pair of scissors. Good luck!

8 Materials crossword

Answer the clues and fill in the crossword puzzle.

Across
2 Iron and steel are both _____
4 It is used to make steel
5 Not a gas or a solid
7 PVC and polythene are examples
8 It doesn't let heat pass through easily

Down
1 Stiff
2 Gold and copper are examples
3 It lets heat pass through easily
6 Oxygen and helium are examples

Reversible and non-reversible changes

The Gang are busy in the kitchen making some food for a party. Zeb is using this opportunity to show the others his knowledge of how some materials can be changed.

Zeb's fact file

A **reversible** change is one in which you can get back the materials you started with.
A **non-reversible** change is one in which the materials are permanently changed.

1 Dotty's fizzy toffee

(a) Zeb is fascinated by Dotty's special toffee recipe because it is an example of a **non-reversible** change, and he takes great delight in explaining this to her.

> Try this recipe whilst thinking about how the mixture of solids and liquid change into a mixture of a solid and a gas.
>
> Start by persuading an adult to help you!
>
> First, lightly grease a baking tin. Then heat four tablespoons of golden syrup and six tablespoons of light brown sugar together in a large pan. Bring to the boil and simmer for five minutes or until the mixture turns dark brown. Now quickly stir in two teaspoons of bicarbonate of soda and you will see the mixture froth up. Pour it into your prepared tin and let it cool completely. When cold, break it into pieces and experience that fizzy toffee taste!

(b) Find out the name of the gas that was released when you added the bicarbonate of soda to the mixture.

2 Milo's icy treats

Zeb's interest has turned to Milo who is making frozen sweets. He explains to Milo that the changes he is making are **reversible**: once the liquids have been frozen, they can be melted back into liquids.

> Try making these yourself.
>
> **You will need** an ice cube tray, a selection of different liquids such as orange juice, yogurt drink and flavoured milk. Almost fill each compartment in the tray with one of the liquids and put them into the freezer for a few hours. When you are ready to eat the treats, take them take out of the freezer and allow to thaw a bit (otherwise you might burn your tongue - honestly!).

3 Heating foods

Zeb shows the Gang how heat can change foods. Some changes are permanent and some can be reversed. Look at the foods on the tray and circle those that will be permanently changed by heat.

4 The water cycle

(a) The water cycle is an everyday example of a reversible change.

Read the stages of the water cycle in the box below. Draw a picture in the box to show the water cycle.

> Water vapour condenses and falls from the clouds as rain.

> The sea is warmed by the sun, evaporates and rises into the clouds.

> Rain collects in the river.

> The river flows into the sea.

(b) Zeb shows Strat how the water cycle works using a pan of hot water.

> If you want to try this at home **ask an adult to help you**.
>
> Put the lid of a pan into the fridge to cool it.
>
> Pour some water into the pan, put it on a cooking ring and heat it until steam starts to rise (**evaporate**).
>
> Hold the cold pan lid a little way above the steam. As the hot steam hits the cold lid of the pan it **condenses** into water droplets and falls back into the pan.

5

(a) **How temperature affects water**

Dotty has put four saucers of water in different places around the house. Write underneath each picture what you predict will happen to the water after several hours. Try to use proper scientific words in your answers.

(i) in the freezer

(ii) in the fridge

(iii) above a radiator

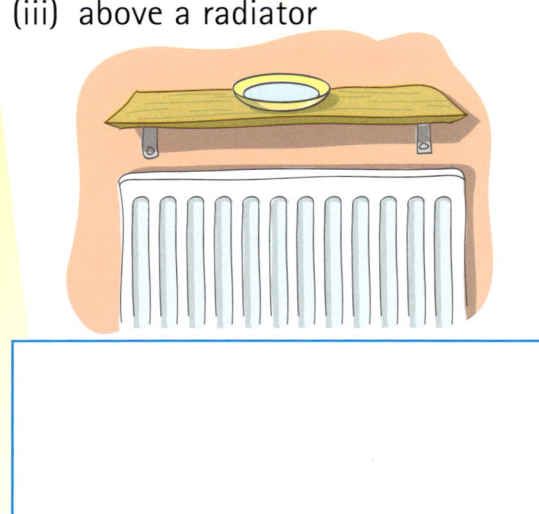

(iv) in a room at room temperature

(b) How could you make sure that each experiment is a fair test?

Write your answer here.

6 More changes to water

(a) Zeb is still investigating the ways in which water changes state. He heated some ice-cold water in a pan. Then he plotted a graph to show how the temperature changed over time.

On the graph, mark with a cross the point where the water starts to boil.

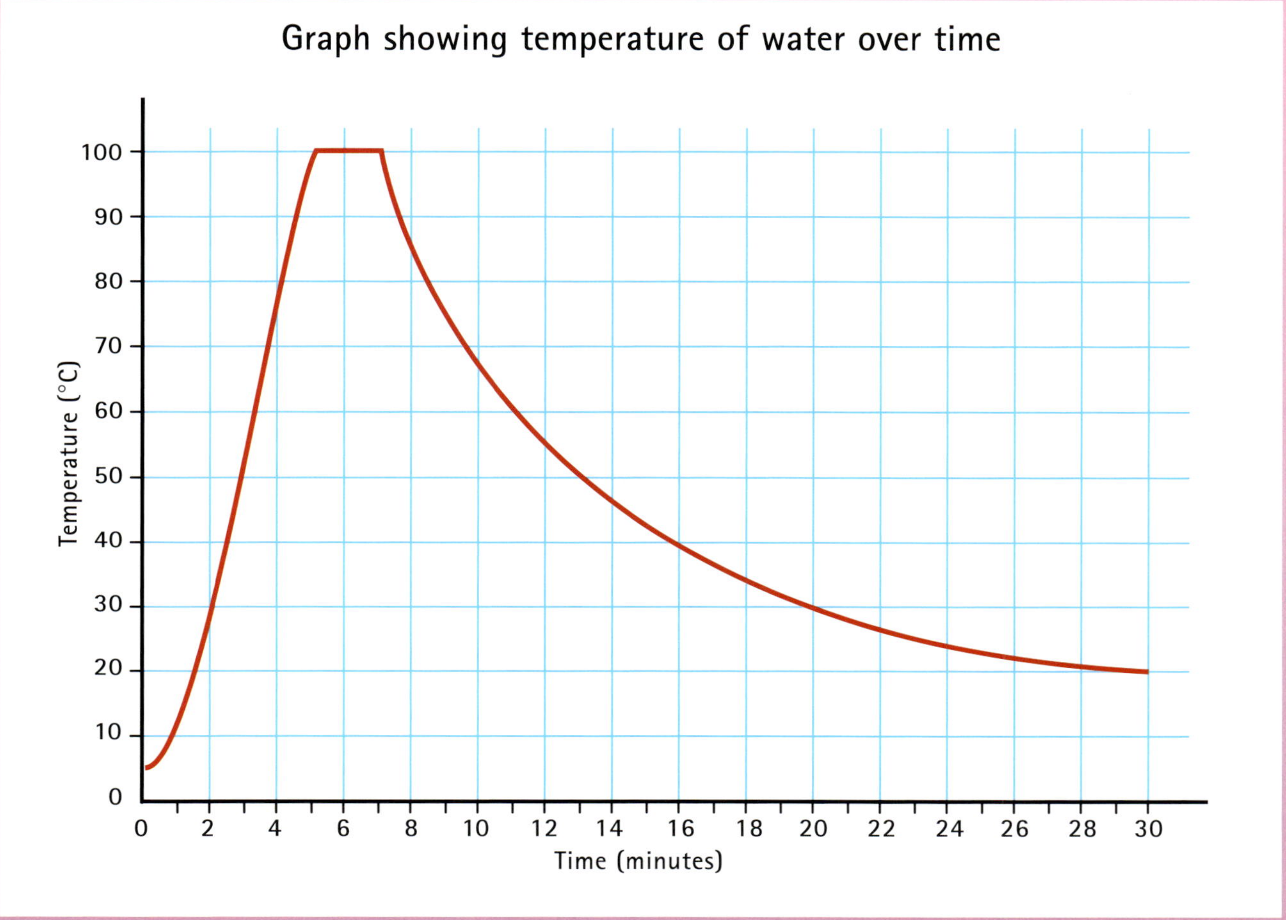

Graph showing temperature of water over time

(b) How long did the water boil for? ☐ minutes

(c) The heat was turned off and the water allowed to cool down to room temperature.

What do you think the temperature of the room was? ☐ °C

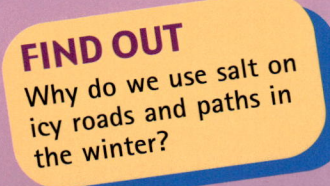

FIND OUT
Why do we use salt on icy roads and paths in the winter?

Separating mixtures

The Elders love raspberry wine but they hate the seeds in it! They have come up with a way of removing the seeds from the wine – simply filter it!

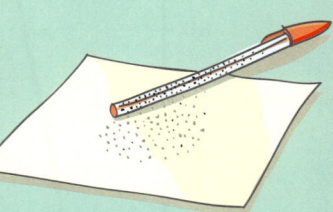

1 Salt 'n pepper

Mix a spoonful of pepper with a spoonful of salt. Challenge a friend to separate the pepper from the salt.

There are two ways to do it.

Method 1
Spread the salt and pepper mixture on a piece of paper. Take a plastic pen and rub it vigorously on a clean wool jumper to create a static charge on the pen. Now place the pen close to the mixture. You will see that the pepper grains stick to the pen.

Method 2
Mix the salt and pepper with water. The salt dissolves but the pepper does not. Now filter the mixture through filter paper (kitchen paper or a coffee filter will do). The grains of pepper will be left on the paper.

The salt is still mixed with the water. You can get the solid salt back by boiling the water to evaporate it; the salt will be left behind. **Ask an adult to help you** with this part.

REMEMBER
A solid dissolved in a liquid makes a solution. Solids which don't dissolve are called insoluble.

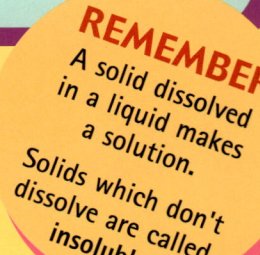

2 Different mixtures

Zeb has concocted a couple of mixtures for you to work out how to separate them.

This mixture contains staples, flour and rice.

This mixture contains mud, stones and water.

Use the words below to complete the sentences.

rice   mud   magnet   filter   stones   sieve   water   flour   sieve

Use a _____ to remove the staples from the mixture to leave flour and rice.

Now use a _____ to separate the _____ from the _____.

Pour the mixture through a _____ to remove the _____ then _____ the mixture to separate the _____ from the _____.

3 Sugar lumps

(a) The Elders have been experimenting to see how quickly a lump of sugar dissolves in a cup of tea. Here is a graph of their results.

DID YOU KNOW?
Teabags are small filters. They allow the tea to flavour the water but keep the leaves inside the bag.

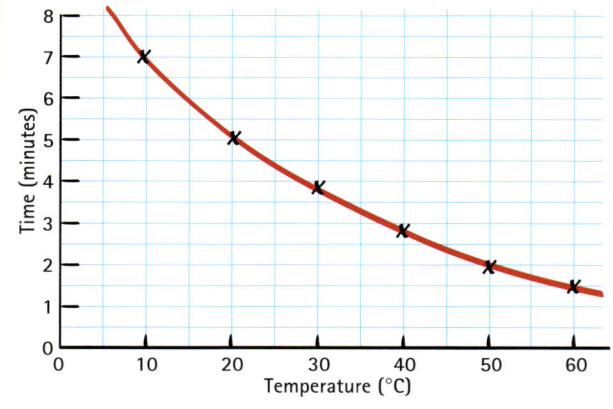

Graph of time for a sugar cube to dissolve compared to temperature of water

How long does it take a lump of sugar to dissolve at 20°C? ☐

(b) At what temperature does it take a lump of sugar two minutes to dissolve? ☐ °C

(c) Describe how the temperature of the tea affects how quickly the sugar lump dissolves.

31

The Earth and beyond

The Phliplids are feeling a little homesick. Some of them were seen gazing up at the stars, looking for their distant galaxy of Phlepland.

1 A solar system wordsearch

Find the words listed in the grid. The planets are listed in the order of distance from the Sun, with Mercury the closest and Pluto the furthest away.

Mercury, Venus, Earth, Mars, Jupiter, Saturn, Uranus, Neptune, Pluto, Orbit, Planet, Moon, Sun

g	e	r	s	j	t	w	s	s	h
m	e	r	c	u	r	y	u	k	i
o	t	u	l	p	n	g	n	u	s
b	o	r	b	i	t	m	a	p	u
r	m	x	s	t	h	k	r	l	n
d	a	o	n	e	p	t	u	n	e
k	r	v	o	r	m	f	r	q	v
u	s	n	h	n	a	c	r	a	d
g	p	l	a	n	e	t	s	a	e
s	m	n	r	u	t	a	s	v	c

2 An orbit challenge

(a) Using these labels, complete the diagram.

Moon   Sun   Earth   28 days   365 days

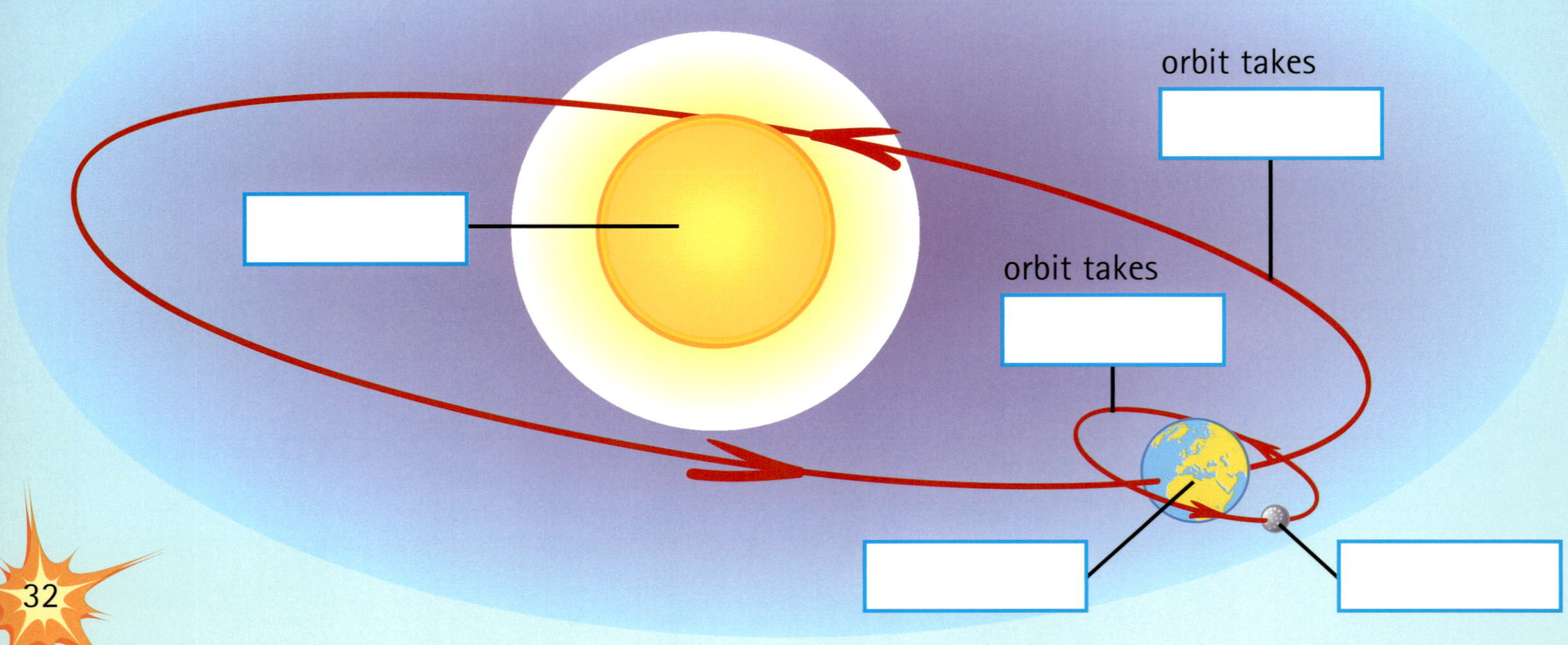

32

(b) The real challenge is to act out the movements of the Sun, Earth and Moon!

You will need 2 friends to help, a beach ball (Sun), 1 pea (Earth) and a tiny bead (Moon). (The ball, pea and bead are approximately the relative sizes of the Sun, Earth and Moon.)

Using your completed diagram from 2(a) on page 32, look at the way the Moon orbits the Earth. As it is doing so, the Earth is orbiting the Sun at the same time. Choose who is to be the Sun, Earth and Moon. Each person finds their 'Sun' (ball) 'Earth' (pea) and 'Moon' (bead) and holds it. Now see if you and your friends can act out the orbits. The 'Sun' person gets the easy job!

3 Moon watch

Astronomy is the study of the movements of the Moon, stars and planets.

Try a little astronomy yourself with this investigation.

The Moon is a sphere, but it doesn't always appear to look like one from the Earth. The shape we see depends on how much of the Moon's surface is lit up by the Sun. Sometimes hardly any of the Moon is seen at all. This is called a 'new Moon'. At other times we see a shape like a crescent, a half or a full circle. The different shapes we see are called **phases**.

Over the next two weeks record the shape of the Moon each night (be patient, you might have to wait for clouds to clear first!).

Fill in the table below.

What's big, bright and silly? A fool moon.

Day 1	Day 2	Day 3	Day 4	Day 5	Day 6	Day 7

Day 8	Day 9	Day 10	Day 11	Day 12	Day 13	Day 14

4 Day and night

Have you ever wondered why each day is divided into daytime and night? This is because the Earth is constantly turning on its axis. Every 24 hours the Earth turns once. When the Sun is shining on the Earth it is daytime and when it isn't, it is night.

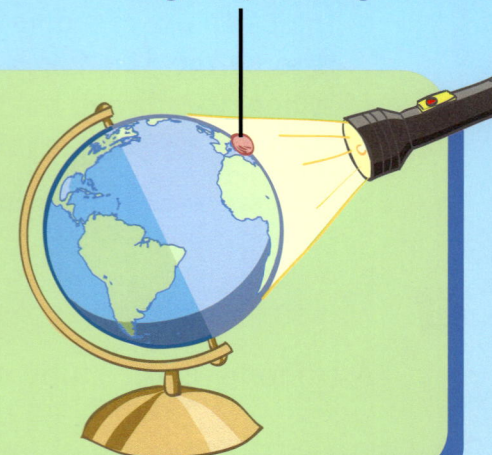

Watch this piece of clay to see how it moves through 'day' and 'night'.

Study this effect using a torch and a globe, or a football if you don't have a globe. You will need to be in a dark room.

Stick a piece of modelling clay on part of the globe. Shine the torch on the clay (day), then slowly turn the globe. The piece of clay will eventually end up in darkness (night), and after a further half turn it will be back in daylight. Obviously whilst some places are experiencing night, others are having their daytime.

5 Make a planets mobile

You will need a large ball of yellow modelling clay, some string, some thread, 9 thin wire rods (you can buy these from craft shops), some thick white card, coloured pencils and a pair of scissors.

On the opposite page are shapes of the planets. They are not drawn exactly in proportion to each other because some of the smaller planets, such as Pluto and Mercury, would be too tiny.

1. Trace the shapes of the planets on to the card. Look in a book or on the Internet to find the colours to use for each one.
2. Cut the planets out. Use a pencil to make a small hole as shown.
3. Cut nine different lengths of thread. Attach one end of the thread to the hole on each planet and the other end to a wire rod (**A**).
4. Cut a length of string and tie a large knot at one end.
5. Mould the clay into a ball around the knot.
6. Push the nine rods into the ball of clay (the Sun).
7. Adjust the positions of the planets along the wires so that the planets are in the correct order from the Sun (**B**).
8. Hang your mobile from the ceiling.

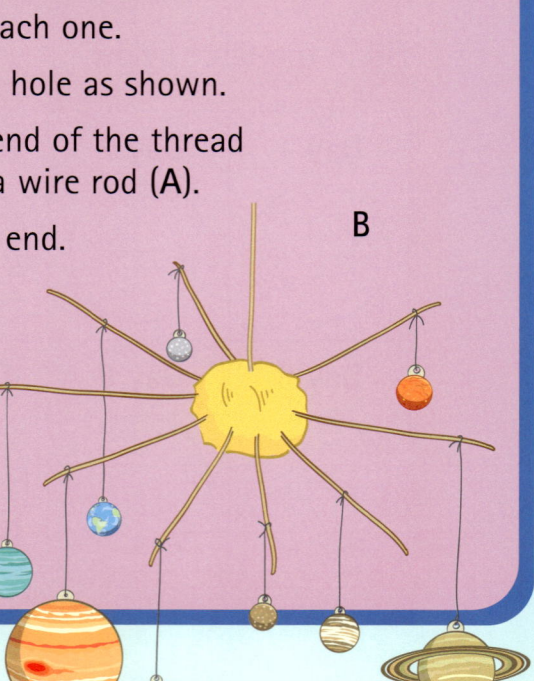

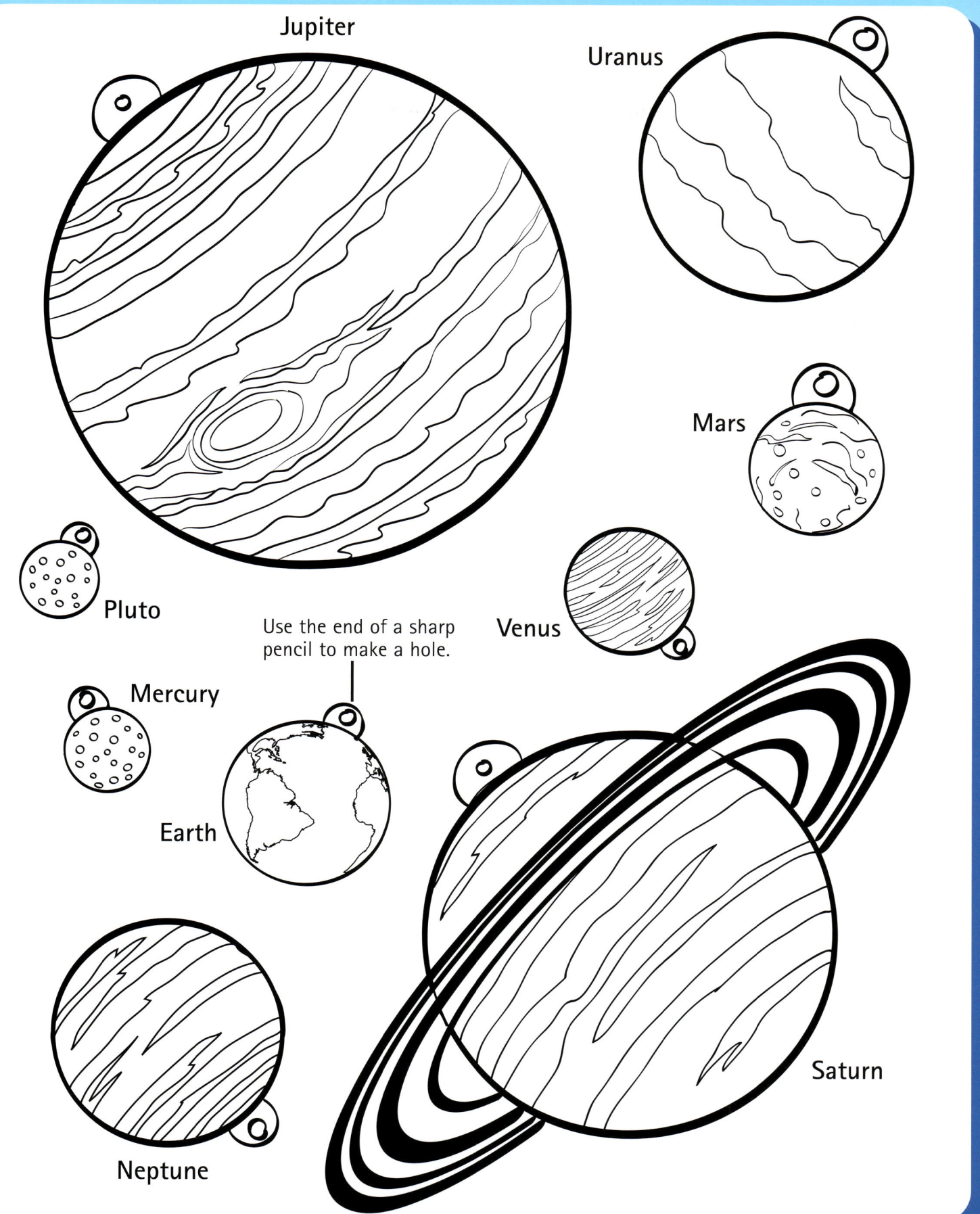

Forces

The Spironauts defy gravity by using jetpacks to get around. The thrust produced by the jet overcomes the force of gravity and allows the Spironauts to fly.

Showing forces

Forces are shown as arrows. The size and direction of the arrow shows the size and direction of the force. These four Spironauts are demonstrating balanced and unbalanced forces using a rope.

balanced forces   unbalanced forces

Gravity and weight

Gravity is the force which acts on all objects pulling them towards the centre of the Earth. There is also gravity on the Moon, but it is not as strong as on the Earth. In space there is no gravity at all.

DID YOU KNOW?
A 60 kg person would weigh about 600 Newtons (N) on the Earth.
A 60 kg person would weigh about 100 N on the Moon.
A 60 kg person would weigh nothing in space!

1 Parachute testing

A parachute is used to slow down a falling object. A force called air resistance pushes the parachute up whilst gravity pulls it down. The parachute keeps falling because the force of gravity is stronger than the force of air resistance.

(a) Draw arrows on the picture of the parachute to show how the forces are acting.

(b) **Make your own parachute**

You will need 2 plastic carrier bags, some thread and some modelling clay. Cut a 20 cm square of plastic. Cut four 50 cm lengths of thread and tie them to the corners of the plastic. Attach the untied ends to the ball of modelling clay (about the size of a marble).

Stand on a chair and drop the parachute. Ask someone to time how long it takes the parachute to fall to the floor.

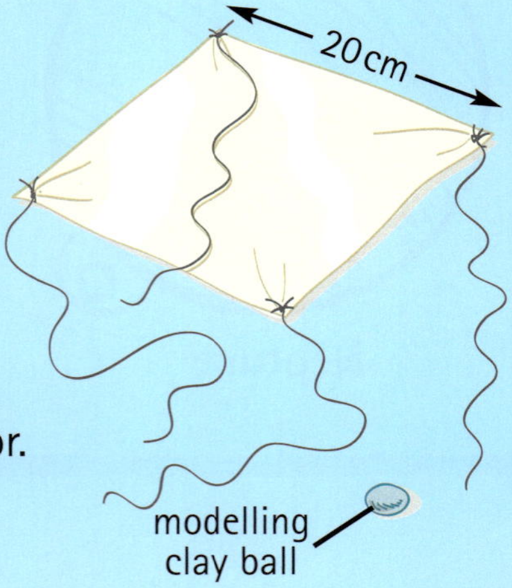

Repeat the experiment, but this time use a parachute made from a 40 cm square of plastic.

To make this a fair test, keep the clay ball the same size and the height it is dropped from the same each time. Record your results.

Size of canopy	Time to fall (seconds)
20 cm square	
40 cm square	

(c) Complete this sentence.

The bigger the parachute the _____ it takes to fall to the ground.

(d) Cut a 10p piece sized hole in the centre of the parachute canopy. Repeat your experiments. What effect does this have?

2 Friction

(a) Friction is the force that occurs when one object slides against another. Try sliding your hands together in opposite directions. You will feel a force and you will also feel heat. Friction makes heat.

(b) In some situations, it is better to avoid friction, whilst in others it is useful. Look at the two pictures below. Describe how friction affects Nina and Strat.

3

(a) **A friction investigation**

You will need a plank of wood about 1 metre long, a toy car, a ruler and different materials to cover the plank. Set up the equipment as shown in the diagram.

Slowly increase the incline of the plank, by adding blocks, until the car rolls down the slope. Measure the height of the highest end of the plank using a ruler.

Now cover the plank with a material, such as plastic, paper, sandpaper, towel or carpet. Measure the height needed to make the car roll down the plank.

Record your results in the table.

(b) Complete the sentence.

> The rougher the surface the _____ the slope of the ramp is needed to allow the car to roll down the plank.

Material	Height (cm)

4 Upthrust

The next time you have a bath take a football in with you. Put the ball in the bath and you will notice that it floats. This is because an upward force called upthrust is balancing the downward force of gravity. Try pushing the ball down into the water and you will feel a force pushing it back up.

5 **A magnet game** (for 2 players)

You will need 2 rods (or pencils), 2 lengths of string, 2 magnets, some paperclips and some sticky labels. Tie the string to the end of each rod and attach a magnet to the other end.

Now stick some sticky labels around some paperclips. On the labels write numbers between 1 and 10.

Scatter the paperclips on a table. Take turns to use your magnetic 'fishing rod' to pick up as many paperclips as you can in each turn. Keep going until all the paperclips have been picked up. Add up the numbers on your clips to see who has picked up the highest score.

DID YOU KNOW?
Magnets and magnetic materials are used in many everyday items.
- Fridge doors use magnetic catches to make sure they shut properly.
- Floppy disks are made of magnetic material and must be kept away from strong magnets.
- The black stripe on a credit card is made of magnetic material.

Zeb's fact file

Unlike poles attract.

| N | S | N | S |

Like poles repel.

| N | S | S | N |

Electricity

The Sparkanon is a huge building in Sparkopolis where lixir is extracted from lixite and converted by a wizzatron into electricity. We have power stations which generate and provide us with electricity.

1 Electricity in the home

In our homes we use lots of electrical equipment. Many of these items use electricity supplied through the mains. We plug them into a socket and switch them on. However, we also use batteries to supply electricity in some items like a torch.

Have a look around your home and make lists of all of the equipment that uses either mains electricity or batteries.

Mains electricity	Batteries
toaster	torch

What happened when the scientist discovered electricity?

He got a nasty shock!

2 A lemon battery

Take a lemon and put two cuts into its surface about 5 cm apart. Put a strip of copper in one cut and a strip of zinc in the other. Put your tongue between the two pieces of metal and you should feel a tingle. This is a very small electric current. Your lemon is working as a very weak battery!

3 Morse code

Morse code is used to transmit messages using electrical signals. Each letter is represented by a series of 'dots' (short pulses) and 'dashes' (longer pulses). Here is the Morse code for the letters of the alphabet.

a	•—	b	—•••	c	—•—•	d	—••
e	•	f	••—•	g	——•	h	••••
i	••	j	•———	k	—•—	l	•—••
m	——	n	—•	o	———	p	•——•
q	——•—	r	•—•	s	•••	t	—
u	••—	v	•••—	w	•——	x	—••—
y	—•——	z	——••				

You can make your own Morse code tapper.

You will need a 4.5 volt battery, some plastic coated wire, a small bulb, (e.g. fairylight bulb, car lamp bulb, torch bulb) and a tapper (see diagram).

Connect your circuit as shown in the diagram.

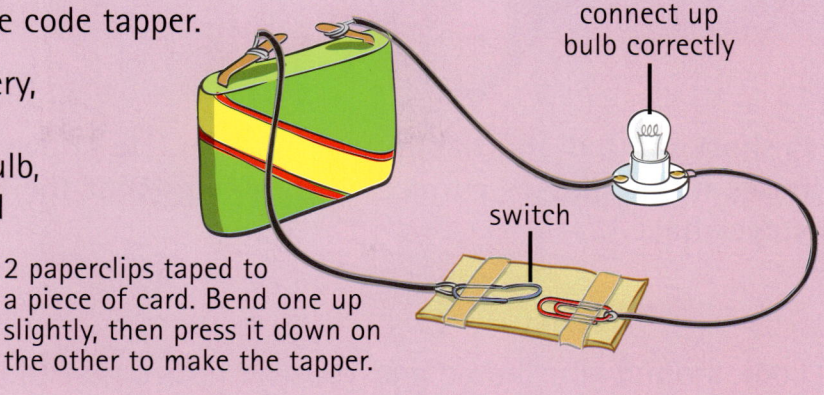

connect up bulb correctly

switch

2 paperclips taped to a piece of card. Bend one up slightly, then press it down on the other to make the tapper.

Make two copies of the Morse code, one for the person sending the coded message and one for the person receiving it, so that the message can be decoded.
Sit at opposite ends of the room. Use the tapper to make short and long flashes of the bulb to match the dots and dashes for each letter. Leave a gap between letters. Can the other person work out your message?

DID YOU KNOW?
There are two ways to make a bulb brighter in a circuit:

(1) shorten the wires

(2) add more batteries.

The first battery was made in 1800 by Alessandro Volta.

The first electrical circuit was made in 1831 by Michael Faraday.

4 Spot the mistakes

(a) The Gang have been testing the lamps in different circuits. But there is something wrong with each circuit. Write the mistake in the box below each picture.

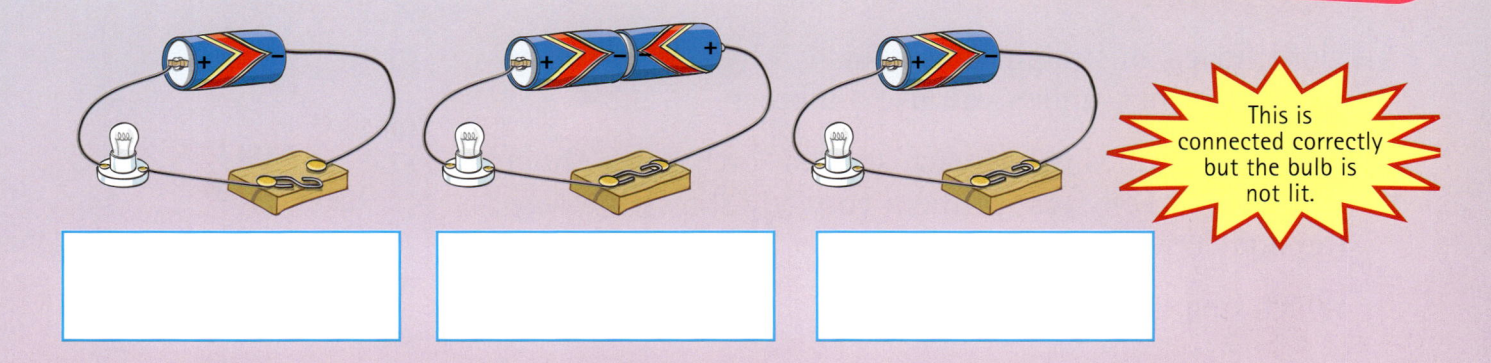

This is connected correctly but the bulb is not lit.

(b) Use these symbols to draw a circuit diagram for the third circuit.

battery

bulb

switch

Light

Zeb is short sighted and has to wear glasses. He knows that the lenses in his glasses correct the problems that the lenses in his eyes have.

1 Light sources

Look around your home and you will notice that there are several objects which provide a **source** of light, and not just light bulbs. There are also objects, such as mirrors, which **reflect** light. Make a list of light sources and light reflectors in and around your home.

Light sources	Light reflectors

What did the light say when it was switched off?

I'm de-lighted.

2 Reflecting light

Light travels in straight lines. This is why you can't see round a corner, unless you use a mirror!

The four pictures below each show a torch being shone on a mirror. Only one of them correctly shows the way the light travels.

Which one is it? ☐

REMEMBER
Transparent: allows all light to pass through.
Translucent: allows some light to pass through.
Opaque: does not let light through.

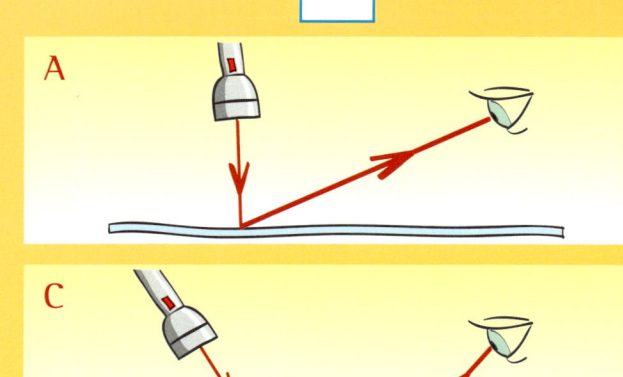

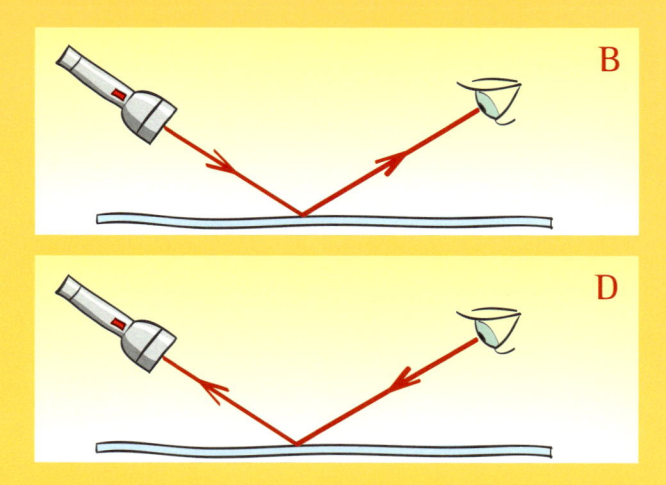

A

B

C

D

3 Multiple reflections

A mirror produces one reflection of an object. What happens if you then reflect this reflection in another mirror? Try this yourself by putting an object in between two mirrors. What do you see?

mirrors

Zeb's fact file

In 1934 Percy Shaw invented 'cats eyes'. These are glass reflectors set into the road to show traffic lanes at night.

4 Silhouettes

When light is blocked by an object, a shadow is formed. Nina is using this to make a silhouette picture of Dotty. Try this yourself.

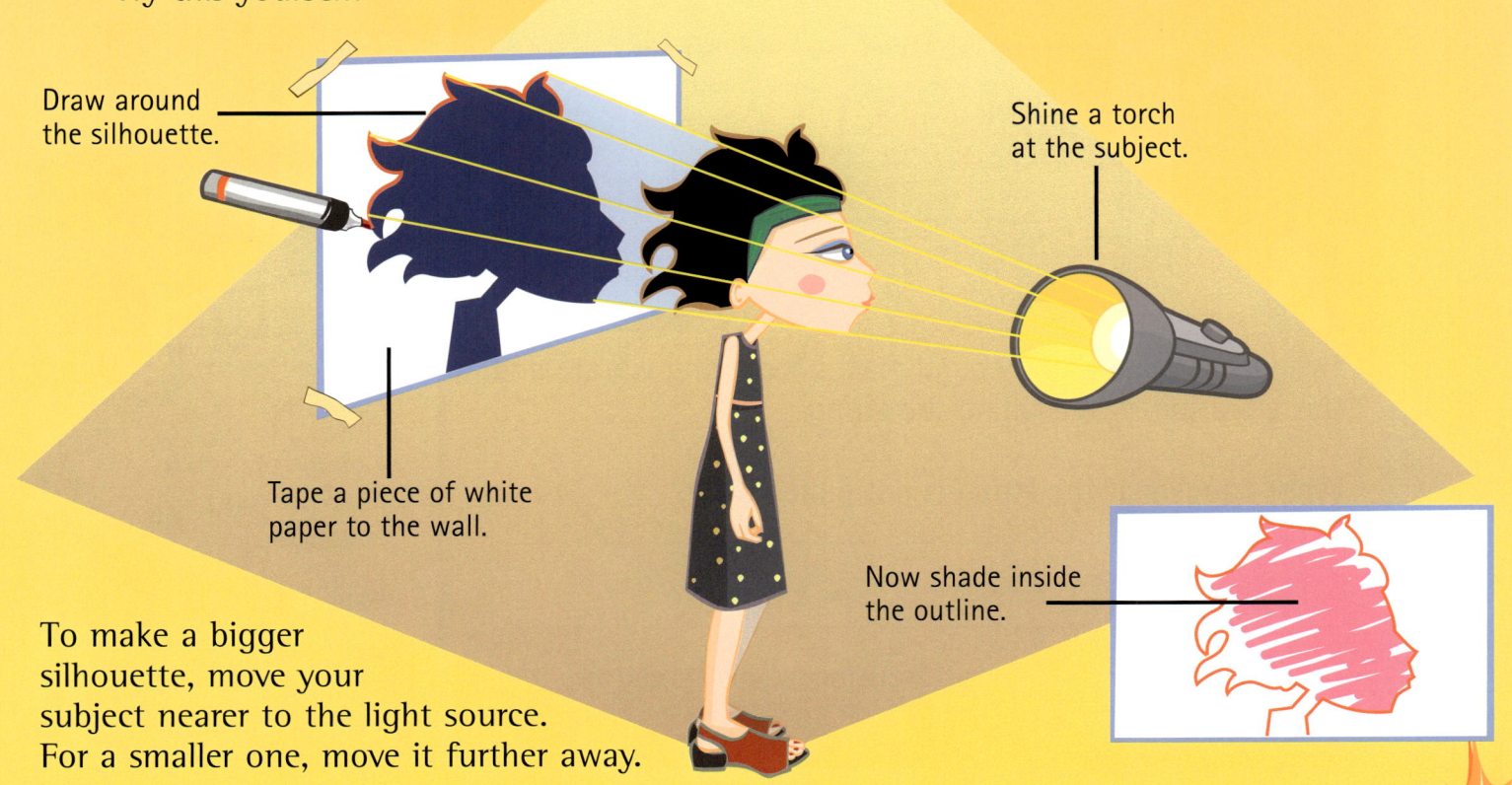

Draw around the silhouette.

Shine a torch at the subject.

Tape a piece of white paper to the wall.

Now shade inside the outline.

To make a bigger silhouette, move your subject nearer to the light source. For a smaller one, move it further away.

Sound

The noisy Spironauts can't hear what anyone says to them as they are always listening to their personal stereos at full volume. Apart from being rather rude, they also risk damaging their ears.

1 Sound is caused by vibration. We hear sounds when these vibrations reach our ears. Sound can travel through different materials.

Investigate how sound travels by making your very own set of cup-phones!

You will need 2 paper cups, about 10 metres of string, fishing line or thin wire and a friend to help you.

Pierce a small hole in the base of each cup. At each end push the string through the hole and knot it inside the cup.

Stand about 10 metres apart and pull the string tight.

Speak into your cup whilst your friend holds the cup to his or her ear and listens. Then you try listening. You should be able to hear what each other is saying.

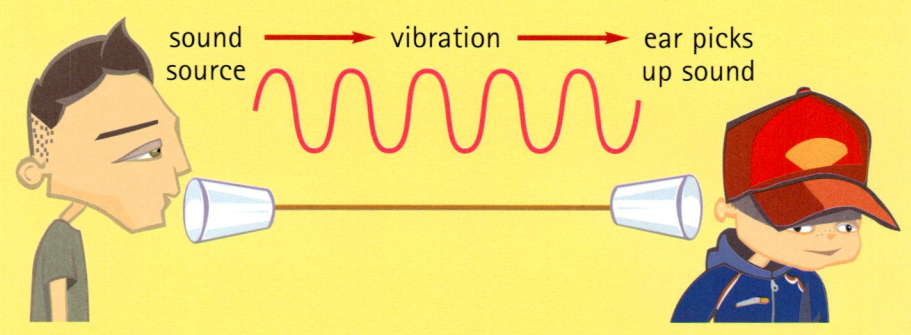

sound source → vibration → ear picks up sound

This works because the sound makes the string vibrate and this vibration travels along the string.

What happens if the string is not pulled tight?

Try repeating this investigation using wool instead of string. Does it work as well?

SOUND BITES

The sound in a quiet room is about 12 decibels, a vacuum cleaner about 70 decibels and a jet fighter taking off is about 130 decibels.

St. Paul's Cathedral in London has a 'whispering gallery'. If someone whispers a word at one side of the gallery, someone standing at the other side can hear it!

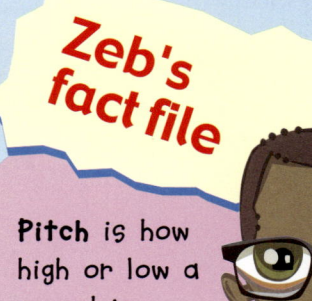

Zeb's fact file

Pitch is how high or low a sound is.

Loudness is how loud or quiet a sound is.

Sounds are caused by vibration.

Sound is measured in decibels.

2 Instruments

Most musical instruments can be plucked, struck or blown. Someone playing a guitar can vary the pitch of the sound by changing the length of the strings being plucked.

Look at the pairs of pictures. For each pair, which instrument makes the higher-pitched sound and which makes the lower-pitched sound? Write high or low in each box.

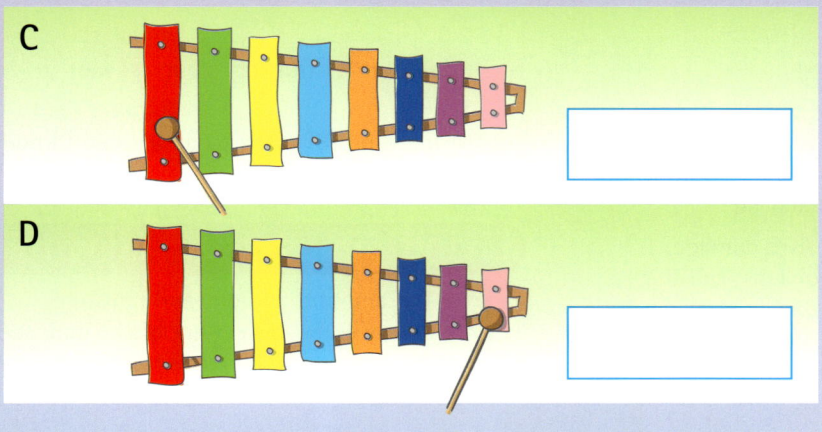

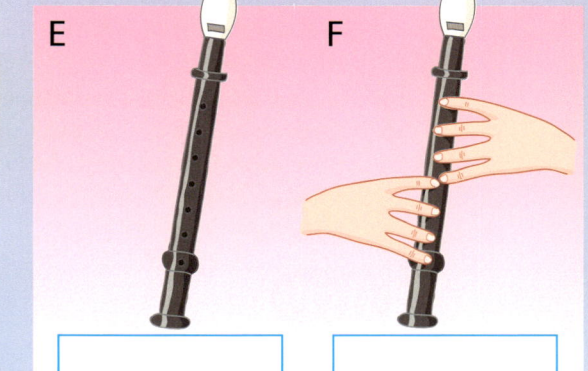

A

B

C

D

E   F

3 A bottle xylophone

(a) **You will need** up to 7 empty glass bottles (milk, wine, beer) or tall glasses all the same size and some water. Put the bottles in a line. Leave the first bottle empty and almost fill the last bottle with water. Pour different amounts of water into the bottles in between as shown in the picture.

Use a pencil to strike each bottle in turn and listen to each note.

(b) Which bottle makes the lowest note? _____

(c) Which bottle makes the highest note? _____

Try making a tune using your bottle xylophone.

4 ECHOES . . . . . ECHOES . . . . . ECHOES . . . . . ECHOES

An echo is a reflection of a sound.

Have you ever noticed that when you shout in a large empty room, such as a school hall or gym your voice echoes? Yet, when you stand in a furnished room, like your bedroom, and shout at the top of your voice you don't get an echo. Why do you think this happens?

45

Answers

Living things and classification

1. excrete, feed, seven, reproduce, sensitive, grow, move
   Spells out: respire
2. (a) sight, touch, smell, hearing
3. field mouse

Healthy lifestyle

1. brown bread
3. brown bread – carbohydrate and fibre; butter – fat; chop – protein; jacket potato – carbohydrate; peas and carrots – fibre and vitamins; apple – fibre and vitamins; biscuit – carbohydrate and fat
6. alcohol abuse – slows reactions, liver damage; smoking – lung cancer, heart problems; sniffing solvents – damages nose, brain damage; taking drugs – risk of overdose, addictive. (In fact all four are addictive.)

Teeth

1. (a) The shell has become soft. (b) The shell is stained purple.

The human body and life cycle

1. *Across:* 2 spine 4 calcium 5 fracture 7 movement 8 heart
   Down: 1 hinge 2 skull 3 muscles 5 femur

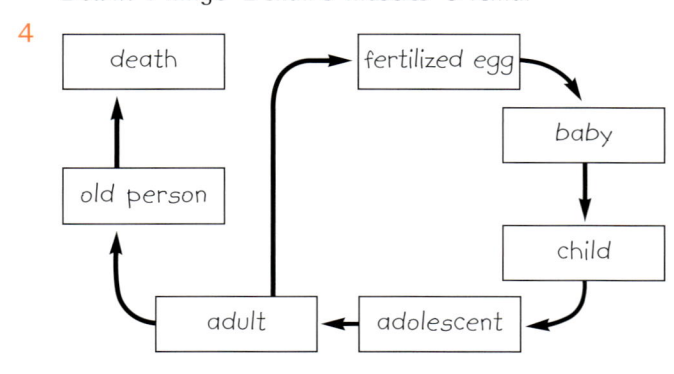

4. fertilized egg → baby → child → adolescent → adult → old person → death; adult → fertilized egg

   Find out: rabbit – 1 month; blue whale – 12 months

Green plants

1. (a) A: The seed germinates. B: A bee visits the poppy. C: The wind scatters the seeds. D: The plant flowers. E: The seed falls to the ground. F: The flower dies leaving a seed head. G: A small plant grows.
   (b) A G D B F C E

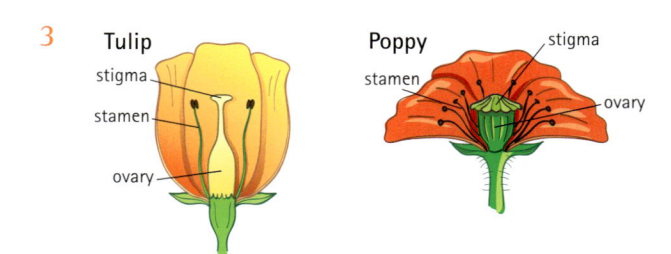

3. Tulip (stigma, stamen, ovary); Poppy (stigma, stamen, ovary)

5. (a) flower (b) fruit (c) stem (d) root (e) leaves (f) root
6. (A) the wind (B) eaten by an animal (C) pods, explode (D) hooks, animals

Living things in their environments

1. (b) The amounts of light and moisture. (c) The woodlice will move to either of the dark sections and eventually they will all move to the dark and damp section. (d) Woodlice prefer dark, damp conditions.
2. forest – squirrel, badger, fern; pond – reeds, frog, heron; seashore – kelp, crab, puffin
4. grass → cow → human; lettuce → caterpillar → sparrow → kestrel; rose → greenfly → ladybird → frog → heron
   (Other combinations are possible.)

Micro-organisms

1. Helpful – yogurt, yeast, compost. Harmful – mould on bread, sneezing, mould on damp wall.
2. newspaper, apple, wool socks, leaves

Properties of materials

1.

	magnetic	non-magnetic	man-made	natural	insulator	conductor
wood		✔		✔	✔	
iron	✔			✔		✔
aluminium		✔		✔		✔
plastic		✔	✔		✔	

2. (a) steel spoon (b) Steel is a good thermal conductor whilst wood and plastic are insulators.
4. solid + liquid = liquid + gas
6. soil with some gravel and sand in it – A; heavy and sticky clay soil – B; very sandy soil – C
8. *Across:* 2 magnetic 4 iron 5 liquid 7 plastic 8 insulator
   Down: 1 rigid 2 metal 3 conductor 6 gas

Reversible and non-reversible changes

1. **(b)** carbon dioxide
3. fried egg, bread
4. **(a)**

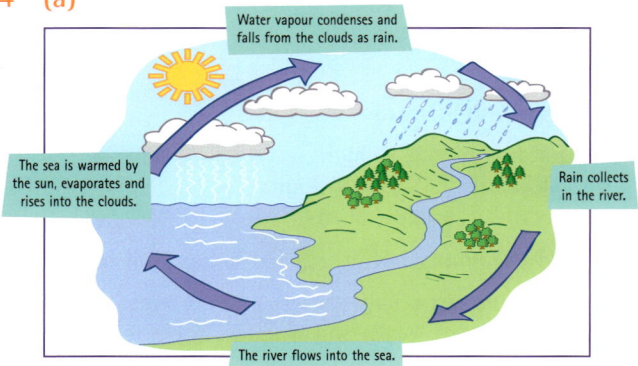

5. **(a)** (i) in the freezer – water will freeze and expand; (ii) in the fridge – water will cool; (iii) above a radiator – water will evaporate completely; (iv) in a room at room temperature – some of the water will evaporate
   (b) Use four identical saucers, use the same amount of water in each saucer and start all four saucers at the same time.
6. **(a)**

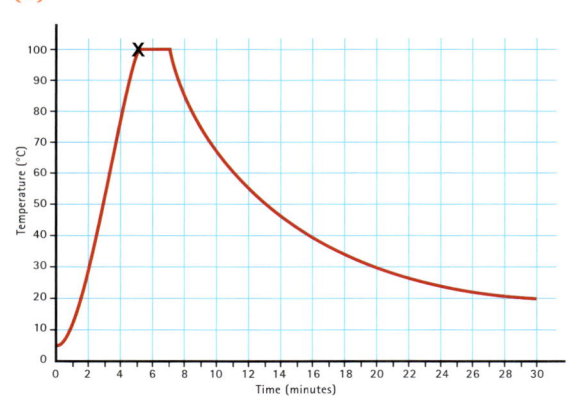

   (b) 2 minutes **(c)** 20°C
   Find out: The salt dissolves into the liquid water in the ice lowering its freezing point.

Separating mixtures

2. magnet, sieve, rice, flour; sieve, stones, filter, mud, water
3. **(a)** 5 minutes **(b)** 50°C **(c)** The hotter the temperature, the more quickly it dissolves.

The Earth and beyond

1.

2. **(a)**

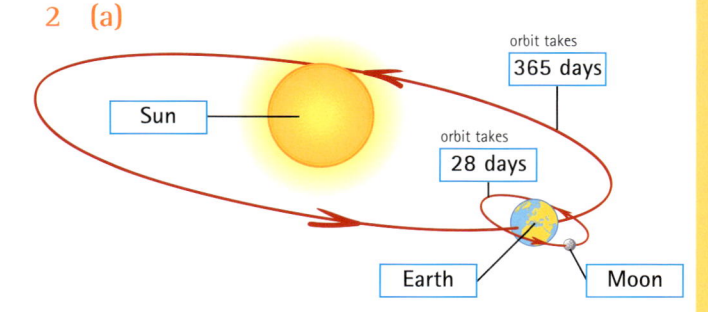

Forces

1 (a)

(c) The bigger the parachute the **longer** it takes to fall to the ground
(d) The hole makes the parachute's fall more controlled.

2 (b) There is very little friction between the skates and the ice, allowing the skater to move easily. The tyres have a heavy tread to provide a strong grip on the ground.

3 (b) The rougher the surface the **steeper** the slope of the ramp is needed to allow the car to roll down the plank.

Electricity

4 (a) (i) the switch is open (ii) one battery is the wrong way round (iii) either the battery is flat, or the bulb is broken

(b)

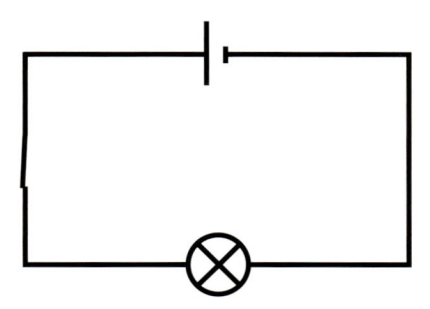

Light

2 Picture B
3 Many reflections of the object are seen in both mirrors.

Sound

1 No sound is heard.
2 guitar – A–low, B–high;
zylophone – C–low, D–high;
recorder – E–high, F–low
3 (b) the empty bottle (c) the almost full bottle
4 Furniture, especially carpets and curtains, absorb sound waves. Walls and wooden floors reflect them.